Cabbage Patch Kids Collectibles

Jan Lindenberger with Dixie McLaughlin

An Unauthorized Handbook and Price Guide

4880 Lower Valley Road, Atglen, PA 19310 USA

Book Designed by Randy L. Hensley
Type set in Windsor BT/Korinna BT

ISBN: 0-7643-0835-1
Printed in China
1 2 3 4

Published by Schiffer Publishing Ltd.
4880 Lower Valley Road
Atglen, PA 19310
Phone: (610) 593-1777; Fax: (610) 593-2002
E-mail: Schifferbk@aol.com
Please visit our web site catalog at
www.schifferbooks.com

In Europe, Schiffer books are distributed by
Bushwood Books
6 Marksbury Avenue Kew Gardens
Surrey TW9 4JF England
Phone: 44 (0)181 392-8585;
Fax: 44 (0)181 392-9876
E-mail: Bushwd@aol.com

This book may be purchased from the publisher.
Include $3.95 for shipping. Please try your bookstore first.
We are interested in hearing from authors with book ideas on related subjects.
You may write for a free printed catalog.

Contents

Acknowledgements

Photos appearing in this book are courtesy of Dixie McLaughlin, Jennifer Bowles and Judy Morris.

I wish to give a special thank you to Jennifer Bowles for her kindness and allowing me to come into her home to photograph her vast collection. Also thanks to Dixie McLaughlin for giving me all the information for this book also for arranging and contributing to the photos. Dixie is an avid collector of Cabbage Patch collectibles and would love to hear from other collectors. Dixie can be reached by writing to her at

Dixie McLaughlin
P.O. Box 133
Lenoir City, TN 37771

Thanks also to Judy Morris from El Paso, Texas. Judy is another avid collector of Cabbage Patch Kids. Mrs. Morris and I are currently working on a Cabbage Patch Kids information and price guide. The book (which may turn out to be two volumes) will contain information only on the Kids. It was wonderful working with Judy.

Cabbage Patch collectors' newsletters:

The Cabbage Line
8500 County Road 21
Clyde, Ohio. 43410
419-547-8367

The Cabbage Connection
610 W. 17th Street
Fremont, Nebraska 68025
acwilhite@teknetwork.com

Prices may differ according to condition, availability, and area. They also tend to vary depending on where they were purchased, whether at a shop, show, or flea market. Most of the prices in this guide are for mint, unopened, in the box, package or on the card. I hope you enjoy this Cabbage Patch price guide as much as I enjoyed doing it. Please use it on all your Cabbage Patch Kids hunts.

Thank you,
Jan

Introduction

Cleveland is a quiet town in the hills of northern Georgia. It is from this area that the phenomenon known as the Cabbage Patch Kids burst upon the unsuspecting world in 1983. Xavier Roberts, a young local artist, began "stitching to life" his soft sculpture babies in 1977, after years of observing and helping his mother make quilts and paying attention to the many and varied crafts produced in the region. His babies were so popular by 1982 that he and the team of "stitchers" he recruited could not keep up with the demand. Xavier named his company Original Appalachian Artworks, and in what proved to be a brilliant move, granted a license in August 1982 to Coleco Industries to produce and mass-market a version of his doll creations with vinyl heads. They were modeled after the original babies. Several companies, including Mattel, Fisher-Price, and Ideal turned down the offer. As they say, "the rest is history!"

As with any highly popular character, a great deal of merchandise sprung up bearing the name and images of the Cabbage Patch Kids. In the early 80s, the advertising agency Schaifer and Nance, having made a deal with Coleco, became the exclusive licensing agent for the Kids. Licensing agreements were granted by OAA for hundreds of items—from clothes, to furniture, to toys, and even food. My mid-1984, there were nearly 70 manufacturers with over 250 Cabbage Patch-related items in the U.S., Canada, and even overseas. Negotiations were completed or underway for almost every country in the world for the distribution or production of the "kids" and their memorabilia.

In May 1984, there was over $200 million in unfilled orders for Cabbage Patch merchandise. The Cabbage Patch Dreams LP record album hit platinum before it was ever advertised. Things eventually settled down and ran smoothly until 1989 when Coleco, in spite of their tremendous success with the Cabbage Patch Kids, fell into financial crisis. Hasbro, the largest American toymaker, bought Coleco in July 1989 and OAA approved the transfer of the Cabbage Patch license for the production of the Kids. During the time that Hasbro had the license, the company made at least 30 new licensing agreements in the UK, France, Germany, and Australia.

In 1994, ironically enough, Mattel, by now the largest toymaker, bid and won the license they had turned down 12 years earlier. They began producing Cabbage Patch Kids in 1995 and their version appeared in stores in Fall of that year. Mattel continues to produce and market the Kids and their related items today.

This book shows just a fraction of the items produced from 1983 until now, and it is my hope that this book will help move Cabbage Patch Kids and their memorabilia further into the collectibles arena.

Happy Collecting!

Cabbage Patch Kitchen

"Jessica Lynn's Birthday Party".
Limited edition plate from
Danbury Mint. 1994. $50 and up

3-piece set by
Wheaton. Loose, $25.
In box, $35 and up.

Royal Worcester 3-piece
porcelain set. 1984.
$50 and up

5-piece ice cream set.
Wheaton. 1984. $25
and up

Back of ice
cream set box.

3 Coke bottles. Left to right:
20th anniversary, $20 and up.
10th anniversary, $35 and up.
15th anniversary, $25 and up

Glass carafe. $2–4

Ceramic mug. $5–7

Back of mug.

Porcelain mugs by Extra Special. $10–20 each

Plastic mug. $2–4

Color-changing (temperature sensitive) plastic cups with lids. Available at Walmart snack shops. $2–4

Child's sipping cup. $4–6

Above- Baby feeding set. $20–25
Left- Plastic cup with handle. $3–5
Below - Left to right: 3-piece melamine
snack set by Wheaton, $25 and up. 3-piece
melamine snack set by Artisan, MIB, $25
and up. Loose melamine plate, $10.

Plastic tea party set. $25–35

Baby feeding set. $20–25

1990s Chilton Party
set. $5–8

Avon toy plastic tea set. $10–15

Pillsbury cake mix with the Hasbro Kids shown. 1994. $20 and up

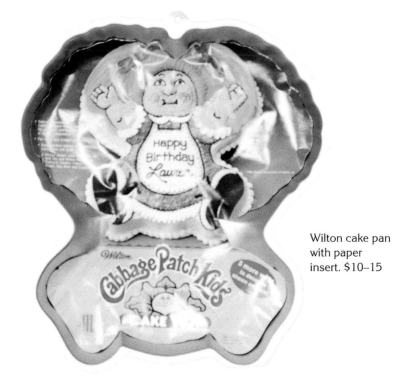

Wilton cake pan with paper insert. $10–15

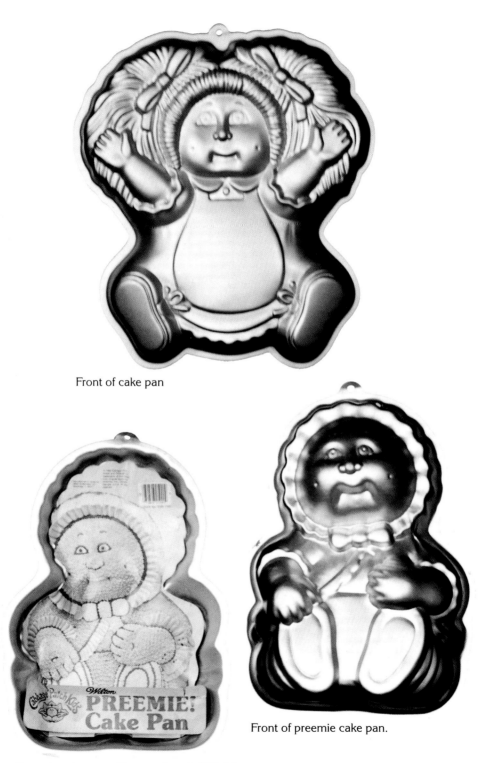

Front of cake pan

Preemie cake pan with paper insert. $10–15

Front of preemie cake pan.

14

Ralston Cabbage Patch Kid cereal boxes. $20 and up

Ralston Cabbage Patch cereal boxes. $20 and up

Ralston cereal.
1985. $20 and up

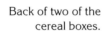

Back of two of the
cereal boxes.

Post cereal from
Canada. 1985.
$35 and up each

16

Back of cereal boxes.

Candy Making Kit.
1984. $15–20

Metal tin, circus
motif. $15–20

Sweet Treasures jelly beans candy
tin. $15–20

Sweet Treasures gumdrop candy
tin. $15–20

Small metal candy tins. $10–15

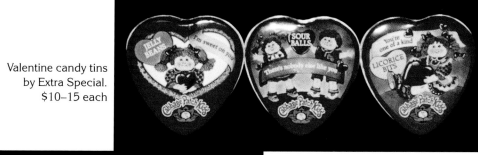

Valentine candy tins by Extra Special. $10–15 each

Real baby bottles. $5–7

Metal lunch box with plastic thermos. $35 and up

Cabbage Patch scene
lunch box, plastic.
Came with yellow
thermos. $7–12

Pink plastic lunch box
and thermos. $15–20

Plastic school
scene lunch box.
$15 and up

Paper lunch bag. $1–2

Metal lap tray. $10–15

TV tray depicting Kids and Koosas climbing mountain. $10–15

Personal Accessories

Cloisonné/metal necklace. Several color variations available. $5–10

Cloisonné necklace. $5–10

Cloisonné/metal pin. $5–10 Several color variations available.

Necklace and
ring set. $4–6

Pin and ring set. $4–6

Ponytail
holders. $4–6

23

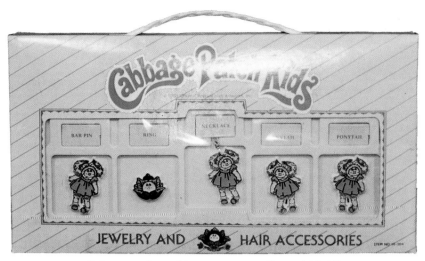

Deluxe jewelry and hair accessory gift set. $10–15

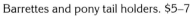

Barrettes and pony tail holders. $5–7

Gold tone Collectors
Club Cabbage Bud
ornament. $15–20

Necklace, earrings, and bracelet by Applause. $10–15 each, mint on card.

Child and doll watch set. $15–20

Pin Me Pretty pin and perfume set. $8–12

Folding comb with
lip gloss. $6–8

Cabbage bud
mirror. $5–7

Avon comb and mirror.
1995. $8–10

Sunglasses, various color schemes. $3–5 each

Mirror and comb
tote. $5–8

Lip gloss. $4–6

Nail polish. $3–6

Comb and mirror set. $10–15

Style Me Pretty comb, brush, mirror set. $20 and up

Color Me Pretty lipstick, nail polish and compact set. $10–15

Fun-in-the-Tub soap set. $15–20

Plastic trinket box. $8–10

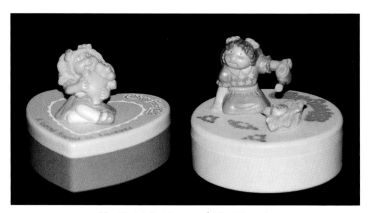

Plastic trinket boxes. $10–15 each

Child's toothbrush set. Missing detachable toothbrushes. $10–15

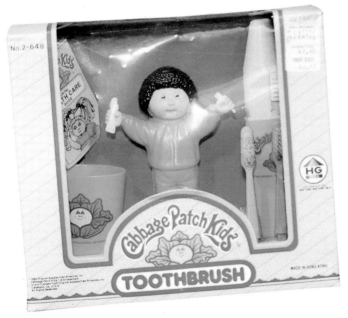

Patches, adhesive strips. Rare. $15–20

Child's toothbrush set. $18–25

Vanity center. $25–30

Light-up vanity set. $30–35

3-piece purse set. Shoulder bag, coin purse, wallet. $10–15 Mint.

Child's plastic wallet. $2–3 loose

Child's canvas wallet. $2–3 loose

Toys & Playsets

Magic catch mitts.
$15–20

Sticker machine. 1985. $15–20

Bubble maker. $5–8

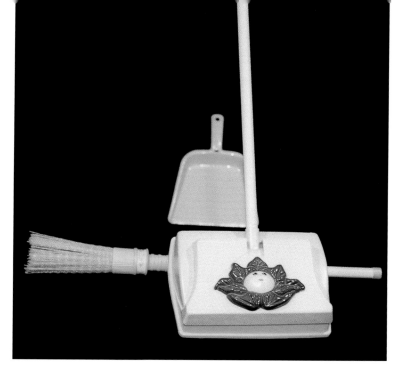

Child's clean-up set. (Not shown: feather duster). $18–25

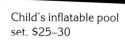

Child's inflatable pool set. $25–30

Sand toy set. 1983. H.B. Ind. $15 and up

Inflatable beach ball. $8–12

Inflatable letter "B" toy. $3–5

Inflatable letter "A" toy. $3–5

Inflatable letter
"C" toy. $3–5

Bunny Bee infant's toy. $5–7

Inflatable toy with bell inside. $3–5

Inflatable toy with bell inside. $3–5

Musical merry-go-round for playmates. 1984. $40–50

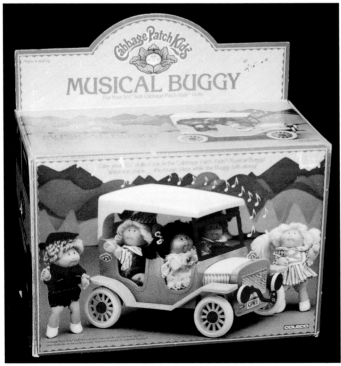

Musical buggy for playmates. $40–50

Playhouse for poseable figurines. $30 and up

Roller coaster. $40–50

Babyland General
Hospital. $35 and up

Babyland General Playset. Comes with 3 figures, boy baby and nurse. $15–20

Babyland General–closed. $15–20

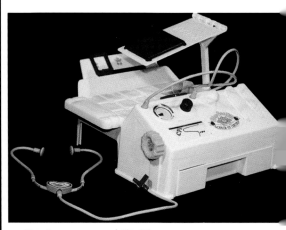

Check–up center. $15–20

40

Activity Sets & Puzzles

Love–N–Go. Babyland school playset. $20 and up

Nursery puzzleforms set. $7–12

Picnic puzzleforms set. $7–12

This is My House sticker book. $5–8

School puzzleforms set. $7–12

42

House stickers for *This is My House* sticker book. 8 different styles. $4–6

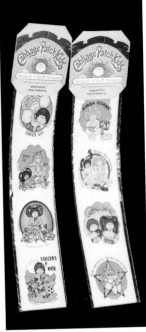

Scratch and sniff stickers. $4–6

Sticker fun book. $4–6

Baby sticker book. $5–8

Poster and pen set. $8–10

Avalon Preemie paper doll. $6–9

Deluxe paper dolls. $10–15

Avalon paper doll. $6–9

Paper doll set. $7–9

Pretty picture paint set. $10–12

Rub n' Play transfer set. $4–6

Sweet Memories Kit. $10–15

ABC Fun coloring book. $3–5

Crayon by number. $4–6

Paint by number. $10–15

1990s Roseart paint by number set. $4–6

Watercolor preemie set. $10–15

Storybook kit. $12–18

Shrinky Dinks. $10–15

Keepsakes kit. $15–20

Suncatchers. $5–9

Deluxe Lite-Catcher kit. $8–12

Go to School kit. $25–30

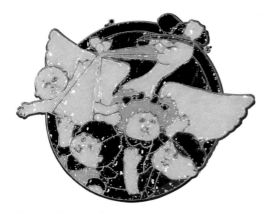

Suncatcher. $4–6

180s Avalon Lite-Catchers. $8–12

Lite Brite re–fill. $10–15

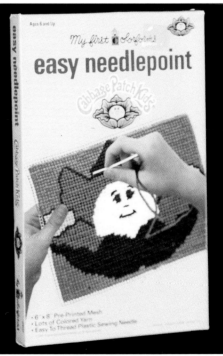

Sweet Sunshine sewing kit. $10–15

1990s Roseart sewing cards. $4–6

Needlepoint kit. $10–15

Stenciling set. $10–12

Cabbage Patch scene puzzle, "Tending the Cabbage Patch". $5–7

Stork stencil kit. $8–10

3D plastic puzzle. $6–8

3D plastic puzzle. $6–8

Magnetic Figure Puzzle set. $10–15

Outside Fun 100-piece puzzle. $5–7

Rock Band 100-piece puzzle. $5–7

Spring Planting 100-piece puzzle. $5–7

Summer Fun 100-piece puzzle. $5–7

Pizza Shop 100-piece puzzle. $5–7

Winter Fun 100-piece puzzle. $5–7

Fall, raking leaves 100-piece puzzle. $5–7

100-piece puzzle. School scene. $5–7

Friends to the Rescue game. $5–8

Woodboard baseball puzzle. $8–10

Large frame tray puzzle. $5–8

Woodboard party puzzle. $6–10

Books, School Supplies, & Calendars

Legend of the Cabbage Patch Kids hardback book. $30–50. Also came in signed leatherbound edition. $100 and up

Little Golden Book. $2–4

Parker Brothers book, *Sybil Sadie*. $5–8

Scholastic book, *Visit the Doctor*. $7–10

Parker Brothers book, *Otis Lee*. $5–8

Parker Brothers books. Left: *Cabbage Patch Kids Adventure*, has photos of kids, $25 and up. Center, paperback activity book, $10 and up. Right, *Xavier's Fantastic Discovery*, hard back, $5–7.

The Big Bicycle Race. Parker Brothers. $4–6

The Shyest Kid in the Patch. Parker Brothers. $4–6

My Own Book. $6–10

Address book. $5–7

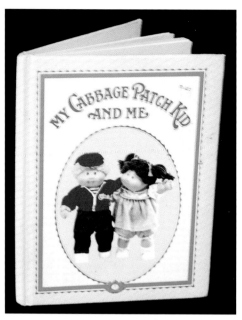

My Cabbage Patch Kid and Me, book. $5–8

Autograph book. $5–7

My Photo Album. $5–7

Crocheted outfits pattern book. $10–15

Hologram stickers. $5–7

Rubber stamps. $4–6

Flat magnets. $6–8

Rubber stamp. $4–6

Small 3-ring notebook. $6–10

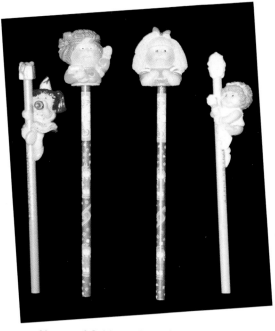

Variety of Cabbage Patch Kids
pencil toppers. $3–6 each

Mini looseleaf binder.
$2–3. $5–8 MIP.

Corkboard sealed in
cellophane. $10–15

Child's chalk-board. $18–22

Pencil topper with pencil sets. $4–6 set

Magic slate drawing board. $4–6

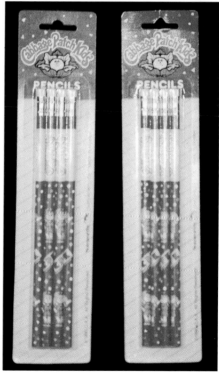

Pencil sets. $4–6

Deluxe stamp set. $10–15

Postalettes by Extra Special. $8–12

Pencil sharpener. $4–6

Spiral bound photo album. $5–10

Spiral bound notebook. $4–6

Pencil tablet. $5–7

School folders. $3–5 each.

School folders. $3–5 each.

Folder. $2–5

1985 calendar. $15 and up

Folder. $2–5

1985 calendar, still sealed. $8–12

Folder. $2–5

1986 calendar. $10–12

1989 Calendar. $10–12

1987 Calendar. $10–12

1988 Calendar. $10–12

1993 calendar. Signed by Xavier. $20 and up

Audio-Visual Items

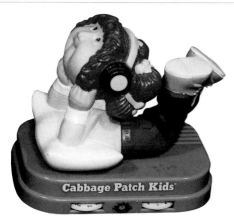 

Cabbage Patch Kids radio. $10–15   Cabbage Patch Kids radio. $10–15

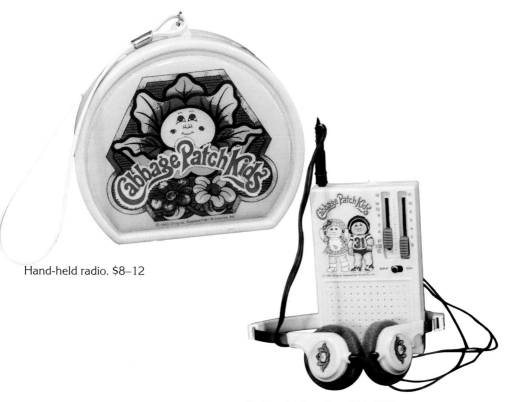

Hand-held radio. $8–12

Radio with headset. $10–15 loose

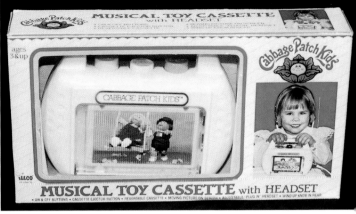

Musical toy cassette. $15–20

Toy wind-up
musical radio.
$10–15

Cassette player. $20–25

Cabbage Patch Dreams
cassette. $6–8

Records and cassettes. Album on left has singing in French. $5–20

Record Tote. $8–12

Record player. $30 and up

Cassette tape case.
$15 and up

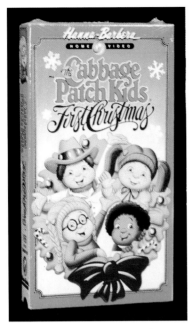

Cabbage Patch Kids' First Christ-mas VHS cartoon video. $15–20

Norma Jean, star of the videos. Sold at Toys-R-Us. $20–25

Real 110 mm CPK camera. $10–15

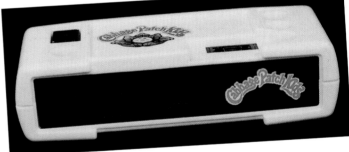

Three CPK videos. $15 each

Holidays and Special Occasions

Large Christmas stockings, various pictures. $20–25 each

71

Large Christmas
stocking. $15–20

Large Christmas
stocking with cuff.
$20–25

Medium Christmas
stockings. $10–15 each

Large Christmas stocking. $18–22

Child's Christmas baby bib. $8–12

Small stockings, $8–12. Large, $18–22

Child's Christmas
bibs. $6–9 each

Winter scene gift wrap. $5–7

Christmas tags. $4–6

Greeting card. $2–4

Greeting cards.
$2–4 each

Christmas ornament in
box. $10–12

Porcelain blonde ornament, hands
down. $10–15

Porcelain blonde girl ornament. $10–15

Porcelain redhead girl ornament, hands up. $10–15

Porcelain orna-
ments, hands
up. $10–15

Coleco 1st birthday card. $1–2

Paper doll wrapping paper. $5–8

Greeting card. $2–4

Summer scene gift wrap. $5–7

3 different wrapping papers. $10–15

Numeral wax candle. $1–2

Colonel Casey gift wrap. $5–7

Party hats. $3–5

Party plates. $2–4

Honeycomb table party decoration. $3–5

Set of six candy cups. $3–5

Party balloons. $5–7

Paper party tablecover. $3–5

Valentines. $3–5

12-foot party
banner. $2–3

Valentines. There was candy in the clear plastic
bubble. $5 and up

Banks

Plastic boy bank that came in a variety of colors. 6" high. $3–6

Plastic banks. $3–5

Plastic girl banks that came in a variety of colors. $3–6

Plastic girl banks that came in a variety of colors. 6" high. $3–6

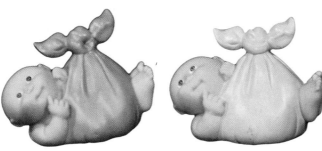

Plastic banks in pink and blue. $3–6

COLLECT THEM ALL !

Back of box showing 4
available banks.

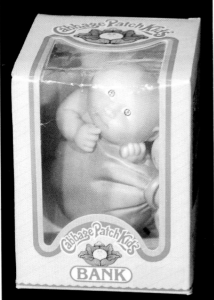

Baby in blanket
bank. Also came
in blue. $6–10

Banks: Cabbage Patch Kid holding heart and
girl in raincoat. $7–10

Plastic bank, boy
with ball. $6–10

Fabulous Fabric Items

Fun-shaped pillow. $10–15

Uncut doll pillow. $5–7

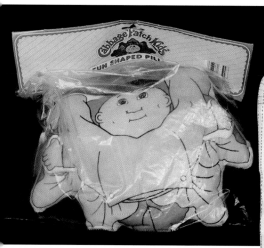

Fun-shaped pillow. $10–15

Uncut doll pillow. $5–7

Uncut doll pillow. $5–7

Uncut doll pillow. $5–7

Doll pillow. $4–6

Doll pillow. $4–6

Doll pillow. $4–6

Handmade pillow. $4–8

Pillow, football boy. $4–6

Comforter. $35 and up

Cabbage Patch Kid comforter. $10–15

Baby blanket. $6–10

Cabbage Patch Kid lace.
$35–50

Child's laundry bag. $15–20

Handmade cross stitch picture. $20–25

Butterick patterns. $5–10

Handmade wall decoration using a
girl with lollipop lace insert. $5–10

Butterick patterns. $5–10

Toddler size 3–D shirt. $15–20 mint

Toddler size 3–D shirt. Bonnet,
hair, pacifier. $15–20 mint

Child's Cabbage
Patch Kid shirt. $3–4

Toddler 3–D (hair, pacifier, bonnet,
bow tie) shirt. $15–20 mint

Child's sandals. $3–5

Assorted children's shoes. 1991.
$20 and up

Mattel doll size in-line skates. $4–6

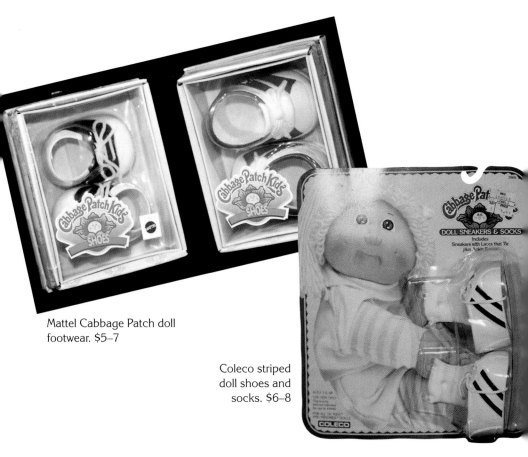

Mattel Cabbage Patch doll
footwear. $5–7

Coleco striped
doll shoes and
socks. $6–8

Mattel doll size party shoes. $4–6

Lion animal sleeper. $20–25

Coleco wired poseable body suits. $4–6

Coleco wired
poseable body
suits. $4–6

One-piece rompers.
$15–20 each

Back of card showing
different outfits.

Hockey uniform set. $15–20

Football uniform. $15–20

Coleco doll's T–shirt. "Why me?" $6–8

Baseball uniform. $15–20

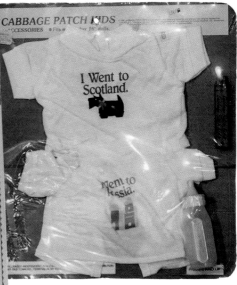

Packaged Cabbage Patch accessories. Believed to be accessory "leftovers" from the final days of Coleco. Items in package could be found with dolls in boxes in previous years. $10–15

Hasbro outfit with special surprise.
When the colorful packet was put in
water, the packaging dissolved to
reveal an accessory for your Cabbage
Patch Kid. $8–12

Mattel raincoat. $6–8

Mattel bedtime outfit. $6–8

Hasbro doll baby
outfit. $5–8

Mattel babies' western fashion. $6–8

1983 Boxed outfit with shoes. $25–30

Boxed raincoat with duck hood. $10–12

Mattel bedtime set. $10–12

Coleco Cabbage Patch diapers. $8–12

Hasbro designer
diapers. $4–6

CPK child-size
slippers. $5–7.
Bunny Bees. Rare.
$25 and up

African-American
Cabbage Patch child's
slippers. $5–8 mint

Brunette Cabbage Patch girl child's slippers. $5–8 Mint

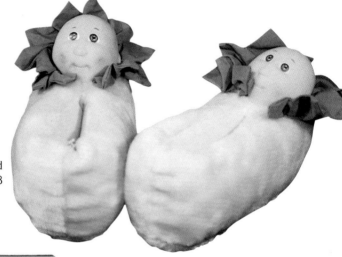

Child's cabbage head slippers. $5–8

Child's play apron. $15–20

Child's slippers. $5–8

Child's elastic waist belt with "head" clasp. $12–15

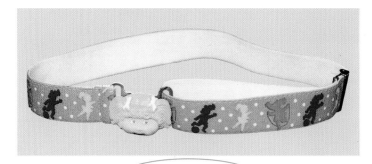

Child's elastic waist belt
with "head" clasp. $12–15

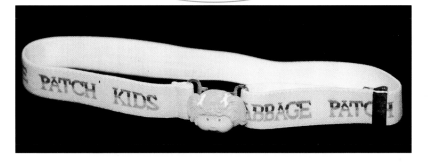

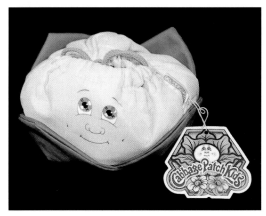

Cabbage head nylon drawstring purse. Also
came in blue and rose. $10–15

Halloween costume. $10–15

Cabbage Patch costume. $5–8 loose

Button-on covers, boy
and girl. $4–6

Button-on covers, roller-skating
girl and baseball boy. Set of 4.
Playskool. $4–6 mint on card

Child's scarf. $8–12

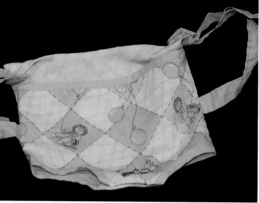

Loose "Kid" carrier. $2–4

Child-size hat, Parents' Association from send-away leaflet in early boxes. $15 and up

Child's bibs. $10–15 each in mint condition

"Kid" soft carrier. $3–5

Cabbage Patch Collector's Club duffel bag. Available to those who were members of the Club from Babyland General Hospital. $10–15

Child's roller skate bag. $15–20

Tote bag with attached change purse. Blonde Kid's head. $10–12

Norma Jean canvas tote bag. $10–12

Child's tote bag. $4–6

1990s. Pink duffel bag. $15 and up

Unique child's 3–D head tote bag. $20–25

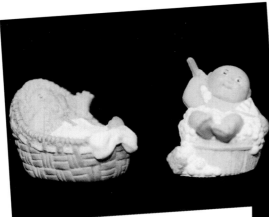

PVC miniatures: baby in basket and boy taking a bath. $3–5

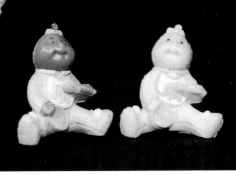

PVC miniatures: babies eating cereal. $3–5

PVC miniatures: baby with blanket and girl working out. $3–5

PVC miniature: baby girl with rattle. $3–5

PVC miniature: baby boy with
bottle. $3–5

PVC miniature: crawling
baby girl. $3–5

PVC miniature: baby girl with
blanket. $3–5

PVC miniatures: baby with hands in air and
girl with flowers. $3–5 each

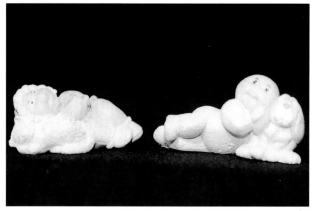

PVC miniatures: girl and boy babies lying with pets. $3–5 each

PVC miniature: little boy doing flip. $3–5

PVC non-poseable figures of baby yawning and girl with ice cream. $3–5 each

PVC miniature: cowboy "kid". $3–5

PVC miniature: little reader. $3–5

PVC miniature: cheerleader "kid". $3–5

PVC miniature: girl in nightgown. $3–5

PVC miniature: baseball "kid". $3–5

PVC miniature: girl wearing rolling skates. $3–5

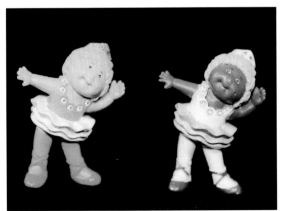

PVC ballerina miniatures. $3–5 each

PVC figures: girl holding heart and girl in raincoat. $3–5 each

PVC miniatures: little girl playing dress up. $3–5

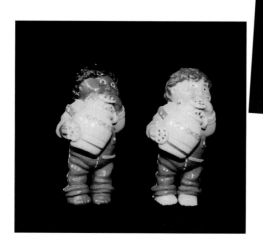

PVC miniatures: little boy eating cookies. $3–5 each

PVC miniatures: birthday girls. $3–5 each

PVC miniatures:
bathtime girls. $3–5

Poseable (head only)
Bedtime figures. $4–7

PVC figurine. $4–6

OlympiKids
1996 figurine
set. $10–15

Poseable (head only) girl figurines. $4–7

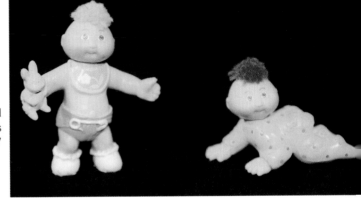

Poseable (head only) babies figures. $5–7

Poseable (head only) cheerleader and pin-wheel figures. $4–7 each

Poseable (head only) Valentine and
Ballerina figures. $4–7

Hasbro poseable figures. $5–8

Poseable Cabbage Patch
Kid figure by Coleco. $6–9
mint on card

Brag bag for Cabbage Patch poseables. $8–12

Rocking baby with wind-up cradle. $10–15

Rocking baby with wind-up horse. $10–15

4 wind-up Kids. Loose. $4–6 each

Cabbage Patch wind-ups: bald boy with freckles and brunette baby. $8–12

Wind-up: Sporting Kid. $5–10

Cabbage head wind-up pop-up. $10–15

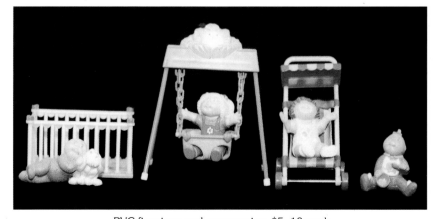

PVC figurines and accessories. $5–10 each

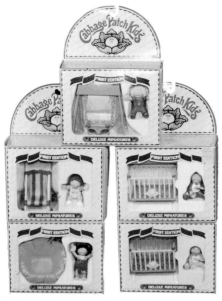

Deluxe miniatures: set of 4. (Playpen set is shown in 2 color variations) $10–15

Playmate. $20–25 Mint in Box

An
ORIGINAL OUTFIT
For Your
Cabbage Patch Kids
Playmate

COLECO

Cabbage Patch Kids
PLAYMATE

OF   IT
Cabbage Patch Kids
Playmate

COLECO

Cabbage Patch Kids
PLAYMATE

Ages 4 and Up

An
ORIGINAL OUTFIT
For Your
Cabbage Patch Kids
Playmate

COLECO

Cabbage Patch Kids
PLAYMATE

An
ORIGINAL OUTFIT
For Your
Cabbage Patch Kids
Playmate

COLECO

Cabbage Patch Kids
PLAYMATE

Cabbage Patch pin-up: Charlene Jenny and her Clubhouse. 6 pin-ups were available; not shown are Toy Shop and Bedroom sets. $7–10 each

Pin-up Brenton Rudy and his barnyard. $12–16 Mint in box.

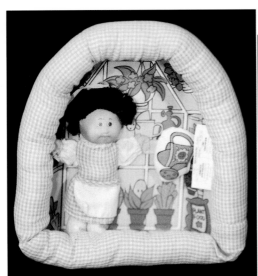

Pin-up Mini Crissy and her greenhouse. $7–10 Loose

Pin-up Candy Jilly and her sweet shop. $7–10 Loose

1990 Mattel small Collectible Kids. $5–8 each. Different hair and clothing combinations available.

Avon playsets. 5" babies. $15 and up

Paint-a-Figurine, baby sitting on a cabbage. $8–10

Paint-a-Figurine sets: boy and girl. $8–10

Justoys Bubble Kids necklace.
Toby Lee. $5–7

Character crayons. $4–6

Justoys Bubble Kids necklace.
Jodi-Ann & Mary-Lou. $5–7 each

1990s pull-back action figures on skate boards. $4–6 each

Porcelain Figurines

These figurines were produced in 1984–1985. Most came boxed and with hang tags. Prices reflect availability and condition.

Porcelain figurine: "Special Delivery". #5063. $50 and up

"Home Sweet Home"
Extra Special club
figurine. $75 and up

Porcelain figurine. Preemie triplets. #5052.
$75 and up

Porcelain music box: "Fun in the Garden".
#5023. $50 and up

Porcelain figurine. Wedding fantasy.
#5062. $50 and up

Porcelain figurine: "Birthday Party". #5021. $15–20

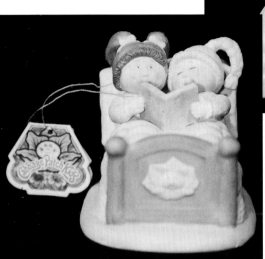

"Bedtime Story" porcelain figurine. $20–30

Porcelain figurine: "Clubhouse". $25–35

Porcelain figurine: "Baby's First Steps". #5047. $18–25

Porcelain figurine: "Be My Sweetheart".
$15–20

"Call me Valentine". $15–20

Porcelain figurine: "Playtime". $15–20

"Tea For Two". #5018. $15–20

Porcelain figurine: Sharing a soda. Figurine is missing a straw in girl's mouth. $18–22

"I Love You". $15–20

"House Call". #5028. $15–20

"Lovely Ladies" porcelain figurine. $40–50

"Getting Acquainted". #5014 $15–20

"Hugs and Kisses". $15–20

"School Sweethearts". #5053. $15–20

"Specially For You". #5005. $8–12

"Findin' Easter Treats".
#5477. $25–30

Porcelain figurines:
"Pillow Talk". $15–20.
"Daydreams" $8–12

"Little Drummer", "Specially for You" and "Best Friends". $8–12 each

"Let's Jazzerize". $20–25

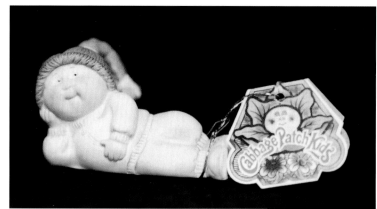

"Waiting Patiently". #5011. $8–12

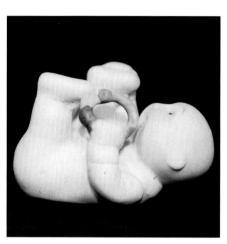

Porcelain figurine by Extra Special; "Playtime". #5004. $8–12

"In Your Easter Bonnet". #5475. $8–12

Mixed group of Cabbage Patch pets. Coleco, 1986, $10–12. Hasbro, 1992, $10–12. Mattel, 1996, $10–12. Coleco bear on right is very rare: $75 and up.

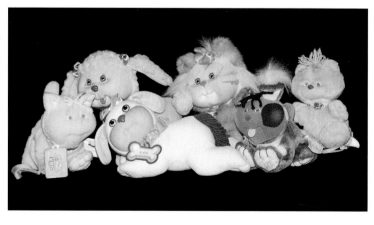

Koosas & Other Furry Friends

The Koosas were introduced in 1984 as critter pets for the Cabbage Patch Kids. They came from Wykoosa Valley, Ga, and arrived boxed with papers and I.D. tag necklace.

Pink and blue Bunnybees. (Also came with shoes). $15–20

Three sizes of Bunny Bees. The one in the center has squeaker in tail. Left to right. 1986, 1994, 1990. $15–25

Tiny Bunny Bees. 1990. Rare. Tag says "Judi's Bunny Bees". $20–25

Hide and Seek Furskins (Sugar, Candy, Honey). $20 and up

Hide and Seek Furskins (Molasses, Jam, Flour). $20 and up

Coleco Cabbage Patch Kid pet dog and cat. $20–25

CPK pet bear with collar. $75 and up Very rare.

Hasbro Crimp and curl Puppy pet. $5–10

Crimp and curl Kitty pet. $5–10

Mattel cat. $10 15

Babyland Collection Rooster. Talks when squeezed. $15 and up

Coleco Show Pony. 1984. $30–35

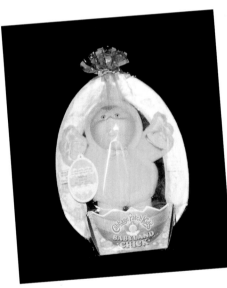

Hasbro Easter basket Babyland Chick. $25 and up

Coleco Circus Pony. 1984. Rare. $50 and up

Lloyd Pluckett chicken. O.A.A. 1994. $40–50

Aunt Nadine Pluckett chicken. O.A.A.
(Original Appalachian Artworks)
1994. $40–50

Vera Mae Pluckette chicken.
O.A.A. 1994. $40–50

Uncle Skeeter Pluckett chicken. O.A.A.
1994. $40–50

Hard to find "eye patch" Koosas. $25 and up

Koosa identification certificate.

Yellow Koosas Cat. $15–20

Koosas Care Center. $20–25

Girl Cat Koosas
Playmate. $20–25

Back of Koosas playmate
box. Shows 6 available styles.

Boy Lion Koosas Playmate. $20–25

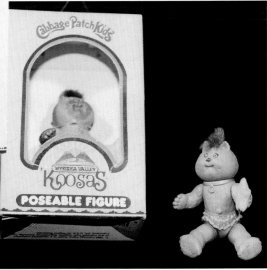

Poseable Koosas figurine. $7–11

Poseable Koosas figurine. $5–7

Koosas poseable PVC figures. $5–7 each

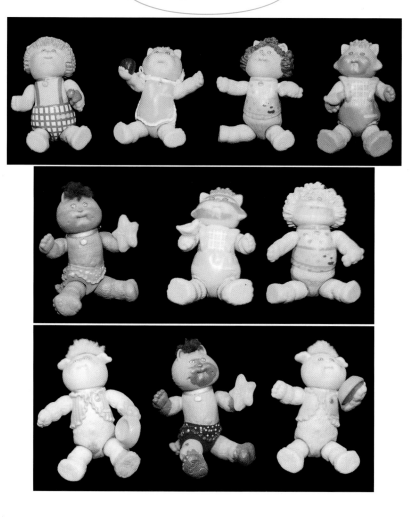

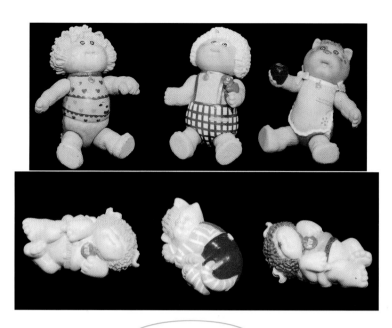

PVC Koosas figurines.
$4–6 each

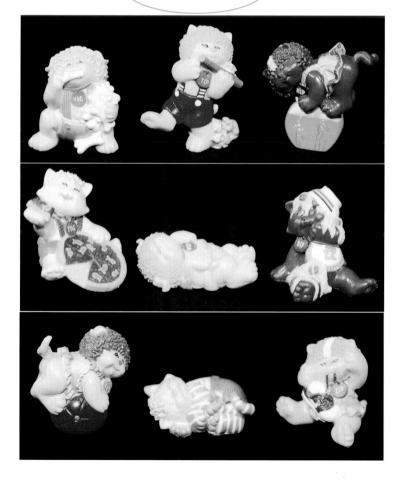

Furskins

The Furskins live in Moody Hollow, GA., Zip code 30528 1/2. The first set was introduced in 1985. They include Hattie, Boone, Dudley and Farrell. The second set arrived in 1986: Fanny Fay, Selma Jean, Jedgar and Orville.

Furskin store display. No price available

First set: Hattie Furskin. $50 and up

Below: First set: Farrell. Approx. 23" tall. $50 and up

Below: First set: Boone as beekeeper and Boone with clothes. $50 and up each

First set: Dudley. Approx. 23" each. 2 different outfits. $50 and up

Second set: Fanny Fay and Selma Jean the Possum queen. $50 and up each

Second set: Approx. 23". Jedgar as Sheriff and Orville as Aviator. $50 and up each

Junie Mae: 16" Coleco Furskin. $30 and up

Bubba: 16" Coleco Furskin. $30 and up

Cecelia: 16" Coleco Furskin. $30 and up

Persimmon: 16" Coleco Furskin. $30 and up

Livingston Clayton: 16" Coleco Furskin.
$30 and up

Hank Spitball: 16" Coleco Furskin.
$30 and up

Lila Claire: 16" Coleco Furskin.
$30 and up

Thistle: Baby Furskin. Also came with
light pink freckles. $35 and up

Furskin book from Wendy's. $6–8

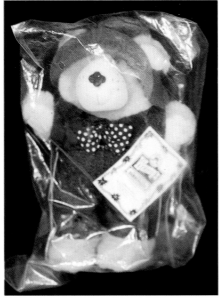

Farrell Furskin from Wendy's. 6" high.
$6–8

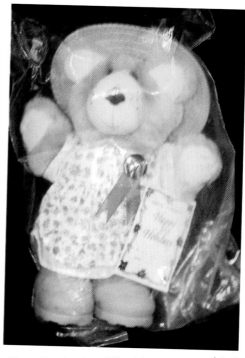

Hattie Furskin from Wendy's restaurant. $6–8

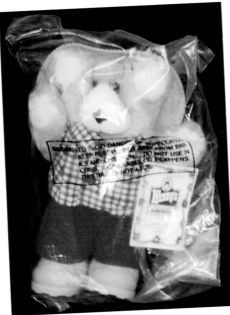

Boone Furskin from Wendy's. $6–8

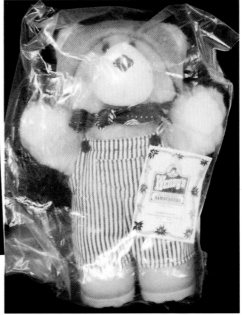

Dudly Furskin from Wendy's
restaurant. $6–8

American Graphics International (incorporated 1988) Jedgar and Orville T. Furskin. 6". $8–10

Dudley and Boone Panosh Place Furskins. 4". $15–20

American Graffics Selma Jean and Fannie Fay Furskin. Identical to Wendy's Furskins but sold in select stores. $8–10 each

Panosh Place Farrell and Hattie Furskins. $15–20

141

Panosh Place Bubba and Cece Furskins. $15–20

Panosh Place Thistle and Hank Spitball Furskins. $15–20

Panosh Place Lila Claire and Persimmon Furskins. $15–20

Panosh Place Jedgar and Fanny Faye Furskins. $15–20

Junie Mae and Scott Furskins. $15–20

PVC Thistle, Persimmon, Junie Mae Furskin miniatures. $5–9 each

Panosh Place Orville T and Selma Jean Furskins. $15–20

PVC Lila Claire and Cece Furskin miniatures. $5–9 each

Furskin miniature. $10–15

PVC Hattie Furskin miniature. $5–9

PVC Scout and Bubba
Furskin miniature. $5–9 each

PVC Hank Spitball and Boone Furskin
miniature. $5–9 each

PVC Farrell and
Dudley Furskin
miniature. $5–9 each

Greeting card. $3–5

Greeting card. $3–5

Furskin greeting card. $3–5

Greeting card. $3–5

Furskin greeting cards. $3–5

Greeting card. Several styles available. $3–5 each

Poster, Xavier with Furskins. $8–10

Furskin calendar. $5–10

Large Christmas
card. $10–15

Boone Furskin Halloween costume. $10–15

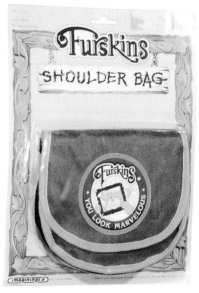

Furskin shoulder bag. $10–15

Furskins bank. $20–25 MIB. $10–15 loose

Child's Furskin tote. $10–15

Furskins travel bag. $15–20

Panosh Place Furskins Moody Hollow general store. $50–60

Furskins battery operated train set. (not shown are 3 PVC figures that came with the set) $20–25

Moody Hollow Express. Furskin train. $35–50

Doll Furniture & Accessories

Rocker carrier for Cabbage Patch dolls. $15–20

CPK stroller modeled by 1985 girl and her "green" friend (not Cabbage Patch). $12–18

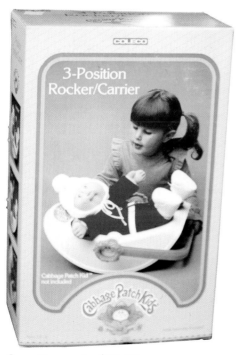

3-position carrier. $15–20

3-in-1 pram. $35–50

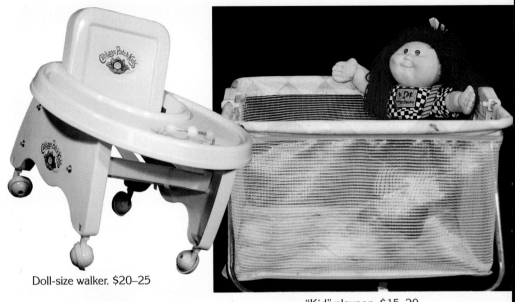

Doll-size walker. $20–25

"Kid" playpen. $15–20

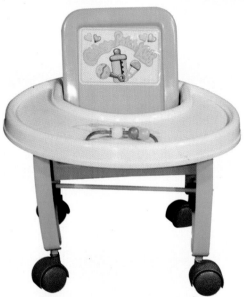

Later version of "Kid" walker. $15–20

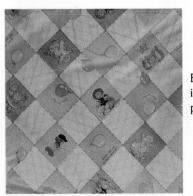

Bottom
inside of
playpen.

Trunk on wheels by Mattel. $8–16

Potty chair. Real noises. 1991.
$15 and up

Fun stroller.
Seat moves up
and down
when pushed.
$50 and up

1990s washing machine. Doll-sized.
$20 and up

Child's ride-on picnic train. Rare. $125 and up

Child's Poppin'
Train riding toy.
$75 and up

Doll-sized tricycle. $25–35

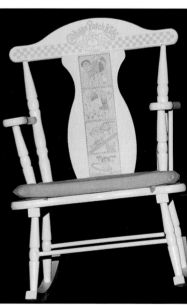

Cabbage Patch Kid wooden high chair. $20–25

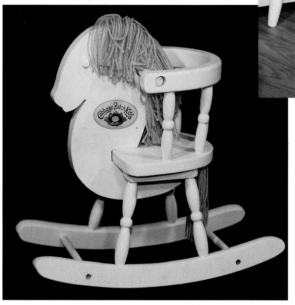

Wooden doll-sized rocking horse. $35 and up

Wooden child-sized rocking chair. $75 and up

Wooden doll cradle.
$30 and up

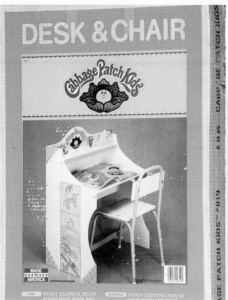

Child-sized desk and chair. $100 and up

Doll-sized storage bench. $35 and up

Metal stand. $20–25

Child's toy box. $25–35

Circus Kid with mustard-colored popcorn hair and tongue sticking out of mouth. $50 and up

Metal waste paper basket, circus motif. $15–20

Metal trash can—different circus design. $15–20

Back of Circus Kid box showing different Kids available

Dutch Holland World Traveler girl with popcorn hair. $40 and up

Coleco cowboy with show pony. $40 and up

Back of box. Shows other countries available: Spain, Holland, Russia, China, Scotland.

Shirley Temple-curl Cornsilk Kid. $30 and up

Real telephone. $75 and up

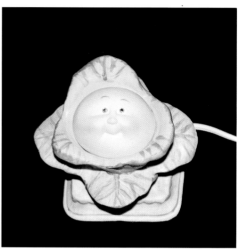

Porcelain Cabbage Patch night light. $20 and up

Cabbage Patch lamp. 1983. Mint $25 and up

Cabbage head night light. $5–8

Child's lullaby night light. $20–25

Send-away Coleco plastic stand. $5–10

Alarm clock. Also came in blue and yellow. $40–50

Metal Cabbage Patch stand. Made for porcelain Kids. $15 and up

Cabbage head made from adult slippers. $2–3

White Cabbage head from Cabbage Patch adoption agency. $50 and up

Assortment of Cabbage bud lapel pins. Large one given for 10 consecutive years convention attendance. Less than 100 made. Head about 2" across, Price unavailable. Others, left to right: Vinyl head, 3/4", $10 and up. All cloth, $75 and up. Came with Patti, the collectors' club baby, $50 and up. Cloth, made in Philippines, $20 and up.

Wallpaper and border. $10 each roll.

OlympiKids pennant. 1996. $15 and up

OlympiKids pennant. $15 and up

Garbage Pail Kids mug. $2–4

Poster. Hand signed by Xavier Roberts. $20–25

McDonald's promotional flyer. 1992. $3–5

Litter bag given at Georgia welcome center. 1992. $5–10

McDonald's Happy Meal display. 1992. $75 and up

1990s OlympiKids hand-held video soccer game. $15–20